EVERY BABY HURTS

A NEW LOOK AT THE RESTRICTIONS AND PAIN THAT BEGIN BEFORE BIRTH - AND RETHINKING WHAT WE CALL "NORMAL" IN EARLY DEVELOPMENT

KENNETH R MOENCH

CONTENTS

ISBN: 979-8-9917058-2-0

Disclaimer:

This book is not intended as a substitute for the medical advice of physicians.

The reader should regularly consult a physician in matters relating to his/her health and particularly with respect to any symptoms that may require diagnosis or medical attention.

DEDICATION

I dedicate this book to my mom, Carol Jean, and to "her boys," (my brothers), Dave, Bob, & Darol. We didn't always have a lot, but we always had each other...and that was enough.

INTRODUCTION

You may not see it. You may not hear it. You definitely don't feel it. But they do. Every baby hurts.

That may sound like a bold claim. And it is. But by the time you finish this book, I believe you'll understand exactly why it needed to be written and why it needs to be read by every healthcare provider, educator, researcher, expectant parent, and grandparent.

Most babies aren't born broken. But nearly all are born hurting in ways the medical world has overlooked for generations.

I didn't know this when I started. I didn't know it when I held my firstborn or even my first grandchild. What I did know was that far too many people, children and adults alike, were walking around in pain, and no one could tell them why.

You've probably seen it yourself:

- The baby who screams for hours every night
- The toddler who won't crawl "the right way"
- The kid who's always fidgeting, falling, or melting down

Maybe someone told you, "They'll grow out of it." Or worse, "That's just how babies are."

I'm here to tell you that is not true.

Over the last thirty years, I've worked with thousands of clients: newborns, kids, adults, and even great-grandparents. Many of them came to me as a last resort. They had seen every specialist, tried every therapy, and exhausted every option. They were desperate, frustrated, and exhausted.

What I began to notice again and again was a common thread. The pain they carried didn't start with a fall, an accident, or an aging body. It started much earlier, before they ever took their first breath. That pain began in the womb, shaped by how they were positioned and compressed before birth.

This wasn't just a theory. It was a pattern I couldn't ignore.

This book is about what I've learned — often the hard way — about how pre-birth posture, tight fascia, and undiagnosed soft tissue restrictions can shape the way a baby moves, feels, and develops. These are not theories. They are observations, backed by results. These are outcomes I've witnessed first-hand and that any parent, therapist, or doctor can learn to recognize and respond to.

I am not a doctor. I am not a researcher in a white coat. I am a guy with strong hands, a deep faith, and an obsession with asking why when someone is in pain.

This book is a flashlight. It offers a way to see what has been hiding in plain sight, not just for your baby but for every person who was once one.

Your baby isn't broken. If they are hurting, you deserve to know why and what you can do about it. Before I could help others understand what I'd seen, I had to wrestle with why I was seeing it and why no one else was talking about it.

I didn't write this book because I hold the credentials most experts trust. I wrote it because I've seen something most professionals are missing: a quiet, physical pain that starts in the soft tissue, often before birth, and stays locked in the body for years. Sometimes it lasts a lifetime.

This book is about recognizing that pain and learning how to release it.

So what am I doing writing a book about pain, babies, and mysteries about the human body that nobody else has considered?

That's a fair question.

The truth is, I'm writing it because this pain is real. I've seen it in thousands of bodies, and once you know where to look, it's impossible to unsee.

Since the 1990s, I've worked with people who had nearly given up. People who had seen every specialist, tried every therapy, and filled every prescription.

Nothing worked. Then they came to us. Something changed.

They stood up straighter.

They stopped hurting.

They started sleeping.

They finally felt like themselves again.

And here is what no one talks about: most of these problems had one thing in common. They didn't start with a car wreck, a bad mattress, or aging. They started in the soft tissue. That soft tissue tension had been there for decades.

Sometimes since birth.

Sometimes since before birth.

This book isn't about theory. It is about patterns I've seen in real people over and over again. It is about what happens when you stop forcing a body into a diagnosis and start asking better questions. It is about refusing to accept that pain is something we just have to live with.

Let me be clear. This book is not anti-doctor. I've worked with incredible physicians, therapists, and chiropractors. But most were trained to manage symptoms, not address root causes. Almost none were taught how to look at fascia or muscle restriction in a newborn.

Yet that is often where the story begins.

If this is all true, you might ask, "Why haven't we heard it before?" "Why don't pediatricians know this?" "Why isn't it taught in schools?"

My answer is simple: our system is not built to look for answers that do not come with a billing code.

Treating tight fascia in a baby does not require medication, surgery, or a machine. It requires trained hands, critical observation, and a willingness to see what is right in front of you.

That does not fit the model. But it works.

I've watched babies go from inconsolable to peaceful in minutes. I've seen kids gain coordination, calm, and confidence in a single session. I've seen parents cry with relief that their child's pain was finally gone.

I've also seen those results dismissed as flukes simply because they didn't come from a lab or a double-blind study.

Here's the thing. Anecdotal evidence is not lesser evidence. It is lived evidence. When the same outcome happens again and again for decades, it is worth paying attention.

What we are doing is, in fact, the scientific method: a systematic approach used to investigate phenomena, acquire new knowledge, and refine or correct previous understandings. Our work with babies is observable, reproducible, and measurable — hallmarks of the scientific method. Pediatricians are documenting the empirical data; we continue to weigh and measure outcomes. This is not just a collection of stories; it is structured observation aligned with science.

So no, I don't have traditional credentials. But I do have thousands of hours of hands-on experience. I've seen what works. And I want to share it with you.

This information might challenge what you've been taught. It might upset some experts. That is okay. My goal is not to win arguments.

My goal is to help people heal and thrive.

If that is your goal too, keep reading. You are in the right place.

CHAPTER ONE
WHY I WROTE THIS BOOK
(AND WHY YOU SHOULD READ IT)

MY JOURNEY

Men make plans. God laughs.

That saying used to make me smile. Now I think it might be one of the most honest things ever said.

My plan was to become a physical therapist. I was enrolled in classes at LSU, UT Austin, and Austin Community College. I was working in rehab clinics during the day and raising two young kids in the middle of it all. I was three classes away from completing my degree. But those final classes were only offered in the middle of the workday, and I couldn't afford to quit my job.

So, I took a detour.

I enrolled in massage school, not because I had any interest in massage (I'd never even had one), but because I needed a flexible schedule. I thought it was just a means to an end.

Turns out, it was the beginning of the path I was supposed to be on.

After I graduated, my boss at St. David's Rehab Center asked me to start doing bodywork with clients. I wasn't sure what to expect but was excited about the opportunity. I used what I'd learned in class and a lot of intuition. I liked deep work—techniques that felt like they were changing something, not just rubbing oil on a sore spot. Pretty soon, the doctors started calling what I did "Marrow Massage" because it felt like I was working all the way down to the bone marrow.

But what shocked me wasn't the nickname. It was the results.

People were getting better, fast. Pain that had been hanging on for months or years was gone in a session or two. Once, I completely straightened a patient's spine that had been twisted by cerebral palsy (temporarily). It changed so significantly that, when she was face down, her PT didn't recognize her. Doctors and therapists were puzzled. I was, too. But I couldn't deny what I was seeing. Clients now show up with crutches, knee braces, or boots, and many walk out without them.

That's when I had to ask myself a hard question: Do I keep chasing a credential that teaches me how to manage or reduce pain, or do I keep doing the work that seems to actually eliminate it?

I chose the latter.

I started seeing clients in the evenings. When I reached one or two clients consistently, I shifted to half days. When that grew to three or four a day, I quit my job. I walked away from my original plan, not because I had failed, but because I had found something better.

And yet, I knew something was still missing.

Over the next 30 years, I kept asking questions. I worked with thousands of people in pain: athletes, seniors, kids. Many had tried medication, surgery, or injections. Nothing had helped.

Eventually, a quiet truth began to form. It wasn't an epiphany all at once. It came slowly, through patterns I kept seeing—pain that didn't line up with the stories I was told. And it came into full view when I treated my granddaughter, Calla Cecelia, just a day and a half after she was born.

That was the moment everything changed.

Because what I realized was this: much of the pain people carry doesn't start in adulthood, or even in childhood. It starts in the womb.

Tight fascia. Compressed tissues. Postural restrictions formed before birth. Fascia is the body's connective tissue. It acts like a thin, stretchy web that wraps around muscles, organs, and bones. When it tightens, especially before birth, it can quietly affect how a baby moves and feels. And we never address it. We don't even look for it. We just wait until the pain becomes unbearable, and then we medicate it.

But what if we didn't wait?

What if we could help babies before the pain turns chronic? What if we could give them a start in life that doesn't include dysfunction, discomfort, or disconnection from their own bodies?

That's why I wrote this book.

I saw something others weren't seeing, and I couldn't stay silent about it.

I've learned more from working with real people than I ever did from textbooks. I've seen lives changed with simple interventions. I've watched exhausted parents weep when their baby finally stops crying, when their toddler finally sleeps, when their child finally walks without pain.

This book is the sum of that journey: years of observation, hands-on experience, prayer, failure, faith, and breakthroughs.

If it gives you even one insight that changes your child's life, or your own, then every detour I took was worth it. But this wasn't something I figured out overnight. It took decades, detours, and more than a few wrong turns.

CHAPTER TWO

THE INVISIBLE WOUND

PAIN THAT BEGINS BEFORE THE FIRST BREATH

Most people think pain starts with an injury: a fall, a strain, an accident. But for many of us, it starts earlier. In the womb.

After decades of working with the human body, I've learned this: when muscle and fascia shorten, they don't release on their own. They stay tight. They stay silent until they speak.

I've seen it again and again in clients from newborns to adults—pain that doesn't align with any known injury. And it all began to make sense the day I treated my grandson Austin.

Let's start there. At the beginning. The beginning of pain.

Many of our clients come to us with pain that's been building for years. They can't remember when it started. Some say, "It's just always been there." Others say, "It's been that way as long as I can remember."

That's not just a figure of speech. It's the truth. For many people, the tightness, dysfunction, or discomfort started before they were even born.

Take my son-in-law, Tom. As a baby, he was labeled "colicky." As a kid and as an adult, he lived with constant leg pain. He thought that was normal. One day, I released his quad, IT band, and hip flexor on his left leg. For the first time in his life, the pain was gone. That's how long his body had been holding onto tension—decades, possibly since the womb.

I had a similar experience myself during a bout of the flu. I was aching all over, and as a bodyworker, I wanted to know why. I wrote down every muscle that hurt. The list was long. A few weeks later, when I was feeling fine, I tested each muscle by hand. Every single one was still tight and sore.

That's when it hit me. The areas that ached matched my pre-birth posture. They also matched the way I slept as a small child: curled up tight, left leg pulled up much higher than the right, just like I must have been in the womb.

Could that be the source? Had I been carrying tightness, pain, and distraction for my entire life without knowing it? Was that true just for me, or for everyone?

I let the thought go. Until I met Austin.

THE FIRST BABY I EVER TREATED

I'd worked with children for years, but Austin, my second grandson, was the youngest person I'd ever worked on. Just three weeks old.

He was beautiful. He was also absolutely miserable during diaper changes. Every time my daughter Lauren or her husband Tom tried to change him, he would scream so loudly it sounded like they were hurting him.

At first, I chalked it up to the usual suspects: diaper rash, circumcision discomfort, cold wipes. But they ruled all of that out. The crying didn't make sense.

Then, at 5 a.m., when I was awakened by those very familiar screams, it clicked. Austin's hip flexors. That had to be it.

He had spent months in a tight fetal position, legs tucked in close. Every time they changed his diaper, his legs were straightened. And every time, those shortened muscles were pulled past their comfort zone.

"I need to work on Austin," I told Lauren. "His hip flexors are probably tight. That's why he's screaming."

Lauren didn't hesitate. "If you think it will help, go for it."

Later that morning, we started. I gently placed Austin in my lap and pressed my middle finger into his left hip flexor. Then I slowly bent his leg up toward his chest and back down again.

He screamed. For three minutes straight.

And then… silence. A deep breath. A pause. Peace.

That leg, once hypersensitive, was suddenly fine. No more pain and no more fussing. I prepared to treat the other leg, expecting more howls. But nothing. The right side wasn't tight at all.

"Of course," Lauren said. "In every ultrasound, his left foot was by his left ear. We could never get a clear image of his face because his foot was covering it. The right leg wasn't pulled up nearly as high."

I asked if we could do a full-body check. They agreed.

I worked through his lower back, glutes, quads, hamstrings, chest, shoulders, arms, and neck. The tighter the area, the more discomfort he showed. But as each part released, he relaxed.

When I finally reached his neck and released the final layer of tension, Austin did something no one had seen him do. He stretched out completely, arms above his head, legs much straighter and relaxed… and fell asleep.

Snoring.

It was the first time his parents had ever seen him comfortable.

WHAT THE ULTRASOUND SHOWED

Austin's 20-week ultrasound confirmed what I suspected. His left leg was tucked tight against his abdomen. His neck was sharply bent. This was not a gentle fetal curl. This was a posture that shaped his muscles, fascia, and comfort from the very start.

After treatment, we took photos. His legs were more relaxed. His body was at ease. Even as a newborn, the difference was dramatic.

We followed up a few months later. He was older, squirmier, and less tolerant of being held still. He cried during the treatment, but the next day, during a quick re-check, there was no pain, no tension, no reaction at all. The release had held.

That first session changed everything. No more screaming at

diaper changes. No more unexplained distress. Just a calm, happy baby with room to grow.

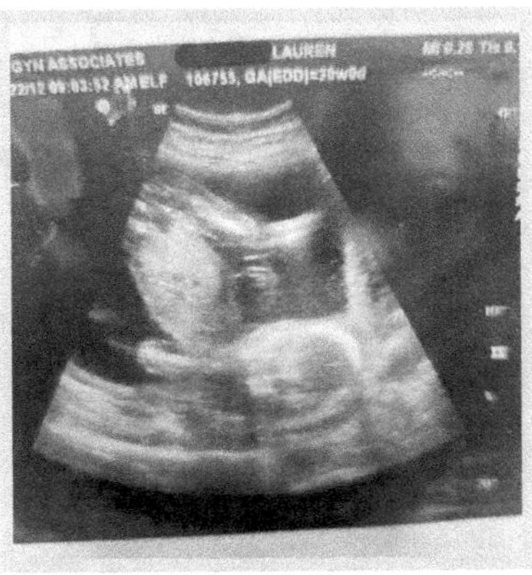

Austin's 20-week ultrasound – with his left leg pulled against his abdomen, and his neck bent at close to a 90-degree angle.

Finishing the last neck release on Austin (in my comfy pajamas).

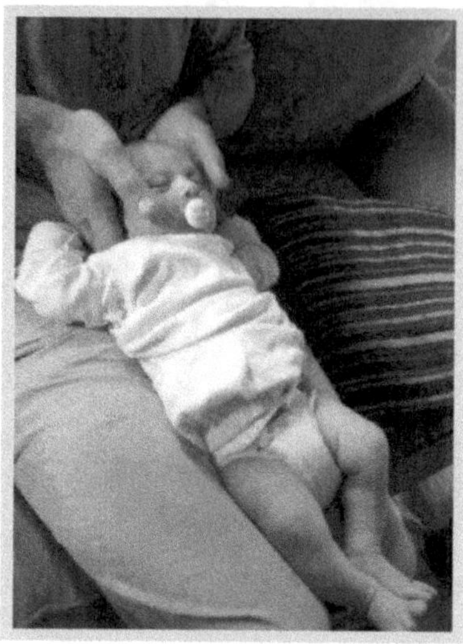

Austin's left leg is still slightly higher than his right leg,
but both legs are much lower than before his treatment.

THEN CAME CALLA

Austin was the beginning. But Calla was the confirmation.

When my granddaughter Calla was born in 2014, I had the chance to work on her just 36 hours after birth.

By contrast, Calla seemed like the picture of perfection: "brand new," "pain-free," and utterly calm. She had ten fingers, ten toes, no visible trauma, and barely made a sound. I visited the day after she was born, and because of Lauren's experience with Austin, she asked when I planned to work on her daughter.

I was curious but not concerned. Calla hadn't cried much and had a perfect Apgar score. The hospital recorded 100 percent oxygenation in her hands, 97 percent in her feet, and no signs of jaundice. She was home less than 24 hours after birth.

Lauren wanted to wait until Calla's pediatrician follow-up, just to confirm everything checked out. It did. The doctor said she was perfectly healthy, so Lauren gave the green light for me to work on her.

Following the same process I used with Austin, I placed Calla in my lap and started at the hips. That's when everything changed. Her hip flexors, quads, lower back, shoulders, and neck were incredibly tight—painful. I was shocked. Though not as severe as Austin's left hip flexor, the level of tension in Calla's tiny body at just 36 hours old was staggering. She fussed and cried through each release, but one by one, her tissues let go.

By the end of the session, less than 30 minutes later, Calla was completely relaxed and sleeping, arms above her head. Just like Austin.

I sat there, stunned. "If this baby hurts," I said aloud, "then every baby hurts." I repeated it, this time after a flood of thoughts, including the heartbreaking reality that some infants are shaken because they will not stop crying.

"Every Baby Hurts!"

It was the moment I realized this wasn't a fluke. It was a pattern. A truth we've missed.

It was a moment of deep clarity. This wasn't a one-time phenomenon. It was a wound we are born with, silently carried until someone has the training and the confidence to help the body release it.

Just like with Austin, the results were immediate and profound. No crying. No fussiness. Just a newborn who was finally free to stretch, breathe, and rest the way her body was meant to.

Lauren later reminded me of how fast her deliveries had been. Her water broke, and ninety minutes later, Austin was born. She pushed just three times. With Calla, she literally fell out. There were no forceps, no hours of pushing, no misshapen heads or cracked clavicles, and no pulling on the baby's head during a C-section. By all accounts, these were considered "easy" births: quick, smooth, and free of intervention.

But that's the part that surprised me. The soft tissue restrictions I felt in both kids weren't the result of delivery trauma. They weren't caused by long labors or mechanical birth injuries. They originated earlier, from the positions the babies held before birth. That kind of in-utero tension can become imprinted in the body before a baby even takes their first breath.

That's the invisible wound.

It doesn't show up on scans. It doesn't always have a name.

But once you see it, you can't unsee it. Pre-Birth Postural Deficits (PBPD) are the muscular and fascial tension patterns in newborns that develop from how they were positioned in the womb.

And once you release it, everything changes.

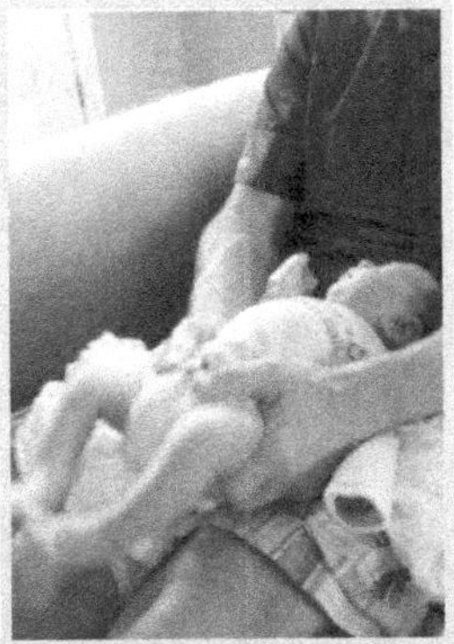

Releasing the soft tissue in Calla's lower back.

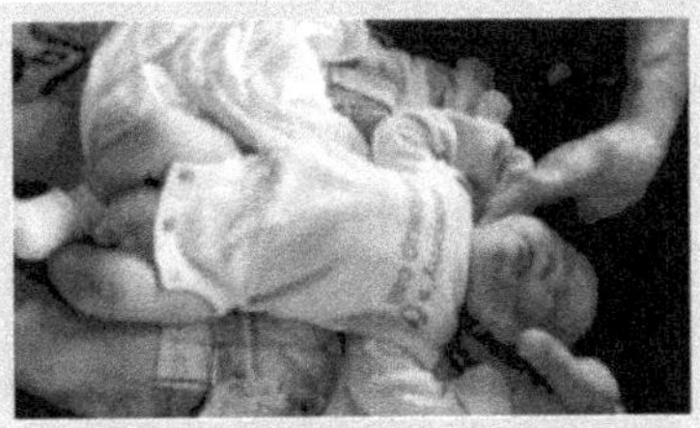

Finishing up with her neck and shoulders.

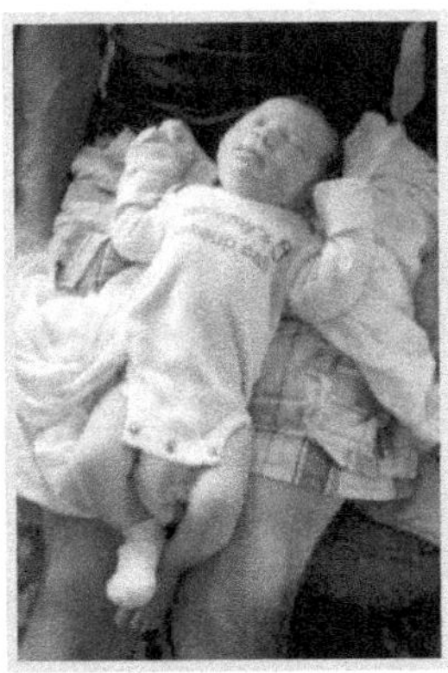

This is one relaxed and peaceful baby... at 36 hours old.

WHAT THIS MEANS FOR YOUR BABY

If you're a parent, you may have noticed your newborn holding one leg tighter than the other. Or arching their back when laid flat. Or wincing slightly when moved. These aren't random quirks. They're signals—signals that something deep in the tissue is still locked from birth or before.

The earlier we address it, the easier it is to release. And the sooner your baby can experience life in a body that feels safe, balanced, and free. Once you've seen the source of the pain, you start to question everything we've been taught about what's "normal."

WHY IS NO ONE TALKING ABOUT THIS?

Why aren't more people talking about this? Why don't pediatricians mention it?

Why don't physical therapists spot it? Why hasn't your doctor ever brought it up?

Those are the questions I hear again and again, usually right after a parent watches their baby stretch, relax, and breathe freely for the first time.

The short answer? No one is looking for it. Not in babies. Not in children. Not even in adults.

It's not part of a routine checkup. It doesn't show up on a scan. It doesn't come with a code the insurance company can bill for.

And because it lives in the soft tissue—where pain hides, where memory stores, where tension builds over time—no one sees it. Pediatricians look for infection. Chiropractors look for alignment. Physical therapists look for functional delays or "weak" muscles. Surgeons look for something to fix surgically. But almost no one looks for patterns of tension rooted in the womb.

This is the blind spot.

And it's bigger than most people realize.

Parents are often told their baby's symptoms are "normal." That they'll "grow out of it." Or worse, that nothing's wrong at all.

But when you've seen what I've seen—when you've placed your finger on a tight hip flexor and watched the crying start, and after only a minute or two watched it replaced with smiles, or released a baby's neck and watched them turn their head for the first time—you stop accepting "normal" as an answer.

This is not about blaming doctors. Some of the most compassionate and capable professionals I know are in medicine. But their training didn't include fascia. It didn't include pre-birth tissue memory. It didn't include the signs that a baby's pain might start before they ever took their first breath.

The system isn't built for this kind of healing. Not because it doesn't care, but because it isn't designed to look here.

But once you see it, you can't unsee it. Once you understand what to look for, the signs are everywhere. And most importantly, once you learn how to release the tension—gently, safely, intentionally—everything can change.

You don't need to be a therapist to start seeing differently. You just need awareness, a little guidance, and the willingness to trust what your eyes, hands, and heart are already noticing.

It's time we pull back the cloak and expose what has been invisible.

It's time we started calling it what it is: real, present, treatable. And it's time we all started talking about it.

CHAPTER THREE
WHY CRYING ISN'T THE PROBLEM
IT'S THE OPENING

Let's be honest. No one enjoys making a baby cry.

But in our practice, "unwrapping" babies is often the most meaningful work we do. In those moments of short-lived discomfort, we unlock the kind of change that transforms a child's life and their family's. Sometimes the shift happens right before your eyes.

What you're about to read challenges much of what we have been taught about babies. These are the common explanations most of us have heard, passed down for generations by well-meaning professionals and parents. What we have discovered, though, tells a very different story.

These are not just theories. They are patterns. And they point to something deeper.

Below is a list of what you may have been told, followed by what we have consistently observed in practice.

Colic: the first "B.S." diagnosis

Torticollis: the second

Tactile Defensiveness: the third

P.U.R.P.L.E. Crying: the fourth

The Standard "B.S." Protocol (used in nearly every culture, often to help parents get some much-needed rest)

Early Chiropractic Care

Tongue Ties

Each of these will be explored in the pages that follow. The goal is not to dismiss them, but to look past the labels. When we understand what is truly going on beneath the surface, we can begin to help babies in a way that is simple, natural, and lasting.

COLIC: THE "WE DON'T KNOW WHAT'S WRONG" DIAGNOSIS

A colicky baby is the "I cry all the time and the doctor has no idea what is wrong with me" condition. When symptoms don't fit a clear diagnosis, many doctors fall back on a familiar label: colic. If a baby cries constantly but tests negative for infection, allergy, or reflux, the default explanation is often, "Your baby is colicky."

The standard advice?

"Change your diet."

"Try a different formula."

"Be patient. It will pass."

Pain is rarely considered. And if it is, the go-to solution is often a prescription for infant or over-the-counter pain medication.

But here is what we have consistently found in practice: when a baby's hip flexors are tight and contracted, they can pull the lumbar spine and hips forward. This shifts posture and compresses the nerves in the lower back. That nerve compression can disrupt bowel and bladder function, leading to a cascade of issues.

The baby's food may not digest properly, which results in gas, bloating, frequent spitting up, abdominal discomfort, and constant crying.

The effects do not always stop in infancy. As the child grows, they might experience chronic bedwetting. Later in life, girls can develop severely painful menstrual cycles. I have treated clients who missed multiple days—sometimes even a full week—each month due to intense cramping. Some were also having 20-day cycles instead of the typical 28-day cycle. In most cases, just one session targeting the hip flexors completely resolved these issues. They have all been incredibly relieved and grateful.

Doctors often tell parents that colic resolves by the time a baby is about three months old. But why three months? That is usually when babies begin moving more: kicking, rolling, and stretching. These movements help lengthen the tight, contracted muscles that have been causing discomfort. As some of the tension eases, so does the crying.

But what if we didn't have to wait three months? What if we addressed the root cause now?

TORTICOLLIS: THE SECOND "B.S." DIAGNOSIS

Torticollis, also known as the "my neck is tight and sore, so I can't or won't move it" condition, affects an estimated 16 percent of live births. That number may actually be higher, as many cases go undiagnosed. The medical literature shows that early intervention is key to avoiding long-term complications.

In fact:

- **Success rates are high.** When treatment begins before six months of age, improvement is seen in 90 to 95 percent of cases. That figure rises to 97 percent when therapy starts even earlier. [*]
- **Recovery is faster with early care.** Many infants respond within just two and a half months of consistent treatment. [†]

What most researchers and clinicians don't realize is this: torticollis can often be corrected in a single session instead of a year of physical therapy.

Let me share a real-world example.

A client of mine had twin boys—Preston and Jasper. Twins are especially prone to Pre-Birth Postural Deficits (PBPDs, which we will dive into later) because of the cramped, awkward positions they endure while sharing the womb.

---

[*] (https://www.ncbi.nlm.nih.gov/books/NBK549778)
[†] (https://www.researchgate.net/publication/272514840_Early_rehabilita tion_treatment_in_newborns_with_congenital_muscular_torticollis)

Both boys were diagnosed with torticollis, and I had the opportunity to work on them. Preston and Jasper were only two months old at the time of this treatment. (The younger the infant, the easier the releases are for the child, the mom, and the therapist.)

Preston's right sternocleidomastoid (SCM) muscle was extremely tight. It pulled his head forward and to the right, which caused a flat spot on the right side of his skull and prevented him from turning his head to the left or even finding a neutral position.

That little SCM muscle put up a fight. But after a few minutes of work, and a few minutes of hard crying, Preston did what so many babies do. He took a few sharp, shallow breaths. Then a deeper one. Then a sigh. And then he let go. His entire body softened.

Suddenly, he could turn his head in any direction. The restriction was gone.

Since he was already on the table, I checked his back, his hip flexors, and other key areas. By the end of the session, he had completely relaxed into the classic post-treatment position we see so often: arms wide, legs long, face soft. Like he was lying on a beach, soaking in the sun. And he was out cold, peacefully asleep.

After his mom picked him up, she couldn't believe the change. He was calm. Alert. At ease. It must have been an enormous relief to finally be free from what must have felt like a tiny straightjacket wrapped around his neck and shoulders.

Jasper's neck was just as tight. And just like his brother, he responded beautifully.

This is the work I get to do every day. And yes, I absolutely love my job.

Jasper and Preston looked uncomfortable BEFORE treatments…

and SO much better after!

Photos from Preston's treatment to correct severe torticollis. This was his first visit, and his condition was completely resolved in just 30 minutes!

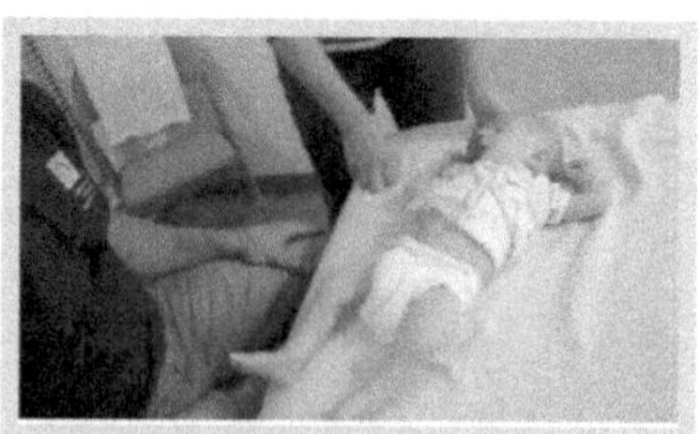

Only head/neck position before RX

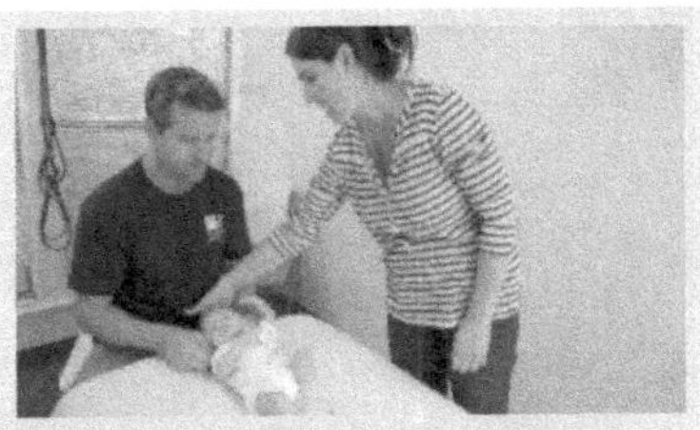

Mom showing flat area on head

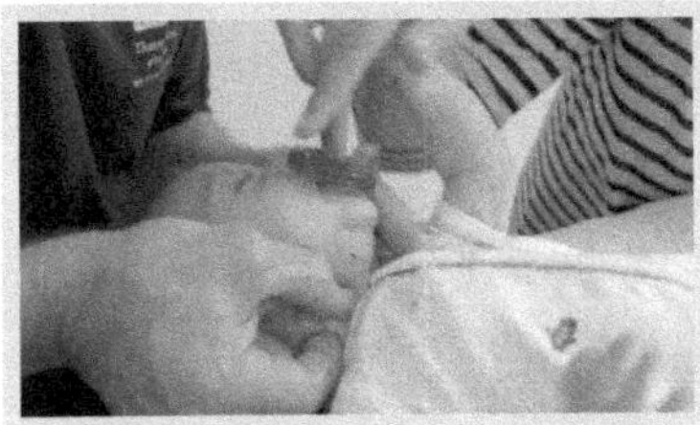

SCM release

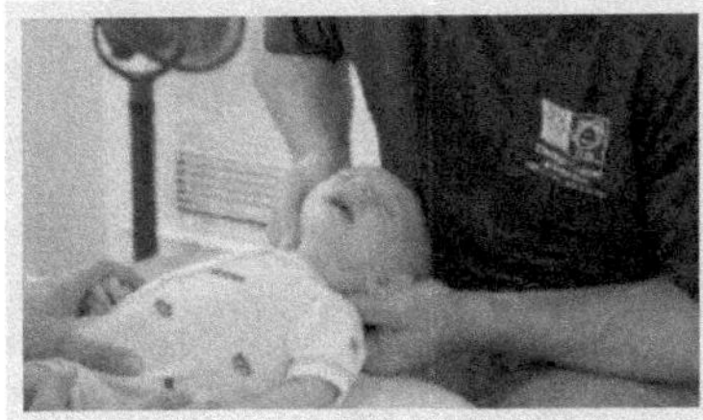

Working other neck muscles.

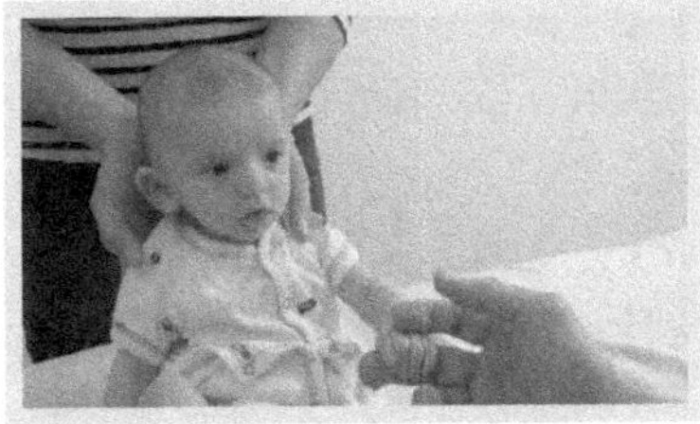

Sitting up with head straight for 1st time!

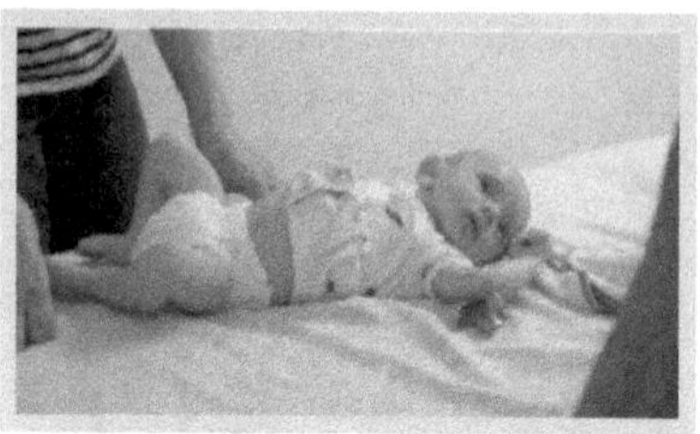

Able to easily turn to the left for the 1st time!

From Claudia, the twin's mom:

"My sons, Preston and Jasper are twins who were both born with severe torticollis. Because their necks were unable to move in other directions the shapes of their heads began changing and their pediatrician recommended baby helmets. This therapy would cost thousands of dollars and was not covered by our insurance. Because my niece's torticollis was cured through Ken's therapy we decided to try that option first. Although the treatment was painful for the babies and hard to watch, I began noticing immediate results and saw their necks turning in ways that they had never been able to before. Before long, the flat places on their heads began changing and the pediatrician was amazed that they were no longer going to need the helmets."

Photos from Preston and Jasper's follow-up sessions at 7 months old—with their mom, aunt, and brother—just to make sure everything was holding!

Discussing twins' position in the womb.

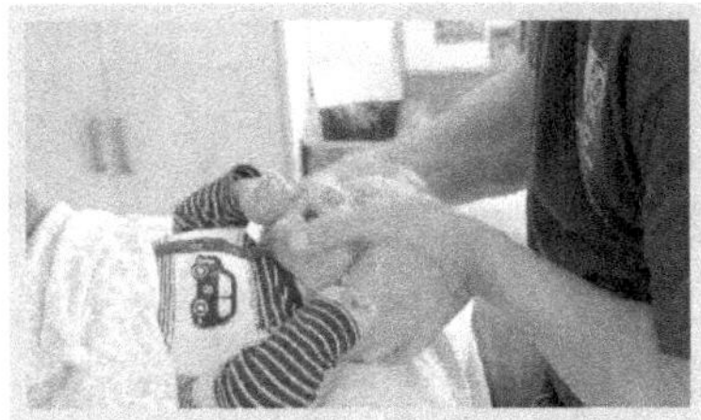

Preston's RX was difficult but effective.

Calming down towards end of treatment.

Treatment is over and everyone is happy!

Twins looking a LOT more comfortable now!

I won't delve into the details of other available treatments, how long they may take, or the potential long-term damage from not addressing this condition properly. The internet is a great resource. Do your research and decide what kind of treatment you'd want for your child or grandchild. As for me, you already know what I'd choose for mine.

TACTILE DEFENSIVENESS: WHEN TOUCH HURTS INSTEAD OF HEALS

This is the "I cry all the time, hate being touched, and no one can explain why" condition.

My older brother Dave was one of those infants who seemed inconsolable. He cried constantly and reacted sharply to even the lightest touch, long before anyone had language for what he was feeling.

Even our family's boxer sensed it. One afternoon, Dave lay on a blanket in the living room, wailing. The dog, desperate for quiet, paced, sighed, shifted position, then finally walked over, stood above Dave's face, and let out a low, steady growl. My parents froze in the doorway, certain he was about to bite.

Instead, Dave went silent. The boxer gave a satisfied huff, padded back to his bed, and fell asleep.

We couldn't explain it then, but the scene revealed how much strain Dave's little body carried, so much that even a dog could feel it and interrupt it. I often wonder how different those early years might have been if the Moench Method had existed: a gentle, intentional way to release the tension no one yet recognized.

Children labeled as "tactile defensive" often flinch, pull away, or become inconsolable when touched. But what if that isn't a behavioral issue? What if it's a pain response?

If a child's muscles are tight and sore all over their body, then every diaper change, every change of clothing, and every loving touch becomes a source of discomfort. That child isn't being difficult. That child is miserable.

And for parents, it's heartbreaking. You want to soothe your baby, to bond with them through cuddles and contact. But they cry harder when you try. Over time, some parents, desperate for sleep and overwhelmed by confusion, begin to distance themselves. Some leave their babies alone in their cribs for long stretches. A few, tragically, lash out in frustration.

These aren't bad parents. These are exhausted, extremely frustrated people who expected joy and instead got screams they couldn't soothe.

And even if the child "outgrows" the label, the consequences may last. Human connection begins with touch. When that's disrupted early, it can delay emotional regulation, interrupt social development, and contribute to challenges that echo far beyond infancy.

Could this be one of the roots of the emotional disconnect we're seeing in older kids and adults? I don't know for sure. But I think it's worth asking how many of these outcomes could have been prevented, if only someone had known how to help that baby's body let go of pain.

It starts with recognizing that this isn't just about sensitivity. It's about suffering.

P.U.R.P.L.E. CRYING – 4TH "B.S." DIAGNOSIS

P.U.R.P.L.E. Crying is an acronym developed by the Shaken Baby Foundation to explain the most common patterns of infant crying. It stands for:

- P – Peak of crying
- U – Unexpected
- R – Resists soothing
- P – Pain-like face
- L – Long lasting
- E – Evening

I have deep respect for the mission and intention behind the Shaken Baby Foundation. Their work is critically important. I would love to work side-by-side with them to help end child abuse perpetrated by parents or caregivers who are pushed beyond their breaking point. But I must be honest: I strongly disagree with the conclusions many professionals have drawn from this model.

The prevailing belief is that babies cry for no reason. That they are not in pain. That it is simply a developmental phase we have to endure.

I believe that these assertions are simply not true.

Babies do not cry for no reason. Neither do adults. We cry because something hurts—physically, emotionally, or both.

The problem is not with the observation. It is with the inter-pretation. P.U.R.P.L.E. accurately describes what parents see, but it completely misses the deeper cause. Instead of asking what might be triggering the distress, the system focuses on managing the symptoms, typically with medications, changes in formula, or sleep strategies.

There is an underlying assumption that infants are inherently flexible and therefore cannot be in pain. Yes, infants have loose ligaments and tendons. But their muscles can still be tight and uncomfortable.

That tension is real. And it is often the source of their crying.

Until the medical community is willing to consider this possi-bility, babies will continue to be misunderstood. And their pain will continue to go unrecognized and unresolved.

I once shared this view in an Instagram post that gained a lot of attention (over eight million views). A doctor from Australia responded, saying that all the research from his institution confirmed that babies are not in pain and cry for no reason. My reply was straightforward:

1. Who funded your research?
2. Can you admit you don't know how to help a baby

stop crying, without actually saying the words, "I don't know how to help a baby stop crying"?

The truth is, most people don't think this work is needed. They see babies as resilient. They assume crying, gas, and fussiness are just part of the package. But just because something is common does not mean it has to be normal. And normal does not mean optimal.

We should not settle for getting through it. We should be striving for something better. Comfort. Health. Relief. The ability to thrive.

That is what we offer every infant we treat.

To any physician, anywhere, I extend an invitation. I will show you, in real time, how this method helps. How it alleviates pain and discomfort in babies, often within minutes.

This is not the time for theoretical debate. This is the time to act. The evidence is there in every single baby we work with. You just have to know how to see it.

SWADDLING – B.S. PROTOCOL

(Used by every culture on Earth... so parents can get some sleep!)

If you're thinking about swaddling your child, I urge you to take a closer look at the research, especially studies that raise concerns about swaddling causing hip dysplasia in some children.

Swaddling is an ancient tradition used in cultures around the world, often with the goal of helping babies sleep better. But does it work the way people believe it does? Not quite.

While many assume swaddling makes babies feel "safe and secure," what it actually does is compress their bodies into the fetal position they were recently in for months. It shortens their muscles, limits natural movement, and stops them from pulling against tight areas that might be causing discomfort. This prevents the kind of stretching and releasing their bodies need in order to lengthen and grow.

The truth is, swaddling can delay physical development. It keeps muscles short and tight, not because it's best for the baby, but because it helps babies and their parents get some sleep.

Some parents try another approach. They let their baby cry in her crib for hours, convinced it's fine. "She just needs to get used to being alone. Crying is normal." This is advice many hear from pediatricians who may not fully understand what is really going on.

But often, that baby is not adjusting. She is in pain.

And tragically, the babies left alone to cry are the lucky ones. Sleep-deprived, overwhelmed, and emotionally drained, some parents or caregivers eventually break. In some heartbreaking cases, they shake or strike a baby who is simply trying to communicate that something hurts.

This is how Shaken Baby Syndrome begins—three words that should never exist in the same sentence. And too often, it starts with pain that went unseen.

Here is the thing. The muscles your baby had in the womb are the same ones they have after birth. Their pain doesn't magically disappear just because they are born.

Imagine putting your arm in a sling for six weeks, with your wrist pulled tight against your chin. Try straightening it afterward. You probably can't, not without weeks of therapy and a slow rebuild of strength and mobility. Even then, you might never get full function back.

Now imagine a baby. Folded up tightly for months. Cramped, twisted, compressed. Then suddenly, pushed through a narrow canal and pulled into the light. No wonder she is crying.

In early pregnancy, babies move freely. But as space runs out, their movement becomes limited. They end up head down, neck muscles shortened, arms folded, wrists curled, legs bent. These positions pull on the lower back and glutes and shorten the hip flexors. Then the birth process begins. Hands reach in, forceps may be used. And out they come, into a bright, loud, overwhelming world.

"Why is she crying so much, doctor?" a worried parent asks.

"Well," the doctor replies, "she has had a rough few hours. You would be crying too."

And if the crying continues?

More tests. More vague answers. Maybe a label like colic. Maybe another formula. Or, "That's normal," and the cycle continues.

Let me be clear. If this book does nothing else, I hope it helps you understand this:

You do not have a "bad" baby. You have a baby who has been confined in a small space for a long time. A baby whose body is trying to stretch out for the first time, and it hurts.

Please find someone who knows how to help.

EARLY CHIROPRACTIC CARE

When it's appropriate… and when it isn't.

Many chiropractors recommend regular adjustments for conditions like torticollis, even in very young infants. In some cases, especially when a vertebra has shifted during delivery, a gentle adjustment may be necessary.

But here is the critical issue: if a baby's muscles and fascia are tight and contracted from their position in the womb, forcing the spine back into alignment can cause more harm than good. When the cervical vertebrae are pushed into place before the surrounding tissues are ready, the already inflamed ligaments and tendons can become further irritated. With repeated adjustments, these structures can be overstretched, which leads to instability.

And when ligaments and tendons no longer hold the spine securely, the muscles are forced to take over. They tighten to provide support, but that compensation creates a new cycle of tension, discomfort, and dysfunction.

This pattern can affect every joint in the body. Ideally, ligaments and tendons offer strong, stable support, while muscles remain flexible to allow dynamic movement. But when connective tissues become too loose, the entire system shifts out of balance. Over the years, some of the most challenging clients I have worked with were those whose spines had been overstretched by frequent chiropractic manipulation. Their muscles were working overtime just to stabilize joints that no longer held themselves in balance.

Chiropractic care can absolutely be beneficial, but only when done correctly and at the right time. In my experience, one thing must happen first: the soft tissue has to be addressed. Muscles and fascia that are tight and contracted must be gently lengthened. Only then can an adjustment hold and deliver lasting relief.

If we see an infant, especially at their second visit, and their neck still does not release fully—meaning they cannot turn their head easily in all directions—we will refer the parents to a trusted Nucca or Atlas chiropractor. We are committed to addressing root causes, not just chasing symptoms. And if the spine is truly out of alignment, that must be corrected before we can achieve a full soft tissue release.

TONGUE TIES: WHY THEY NEED TO BE ADDRESSED

I'm slipping this section in here because it is so important to your child's development and their future well-being, and needs to be addressed by a qualified professional as soon as possible after birth. Austin, Calla, Lauren, and Tom all had tongue ties, so I'm going to let Lauren tell their story. Very cool stuff that I had no idea was even a thing back then.

(In the Central Texas area, we recommend the group "Latched Beginnings" for everything related to tongue tie issues. They are awesome!)

Tongue Ties in Babies and Children

As with the other sections in this book, this information is intended to help identify possible reasons for a child's discomfort. When searching for a diagnosis, Google can be a double-edged sword, offering both plausible explanations and extreme possibilities in the same sitting. While a trip to the pediatrician is often recommended, it isn't always the most productive experience.

For instance, when Calla was about four weeks old, I mentioned to her doctor that she occasionally projectile vomited after nursing. One time, she was nursing on my left side and vomited with such force that it arced completely over my body and down the side of my bed on the right. The doctor advised me to monitor when it happened and mentioned the possibility of pyloric stenosis, a condition that could require surgery. Later, I discovered that the issue was due to her posterior tongue tie, which caused her to take in excessive air while nursing.

This isn't meant as a criticism of our pediatrician—sometimes they simply don't have enough information, as I'll explain later.

I honestly can't even recall how the phrase "tongue tie" first came to mind. It was an obscure post on Facebook that caught my attention with a title about breast milk supply. At the time, Calla was six months old, and I felt like my supply was starting to dip. For context, my grandfather had just passed away. This is relevant because when Austin was three months old, my grandmother passed away, and my milk supply completely dried up within a week—despite having overproduced for the first three months.

I carried a lot of mom guilt over letting stress affect my body with Austin (or at least thinking that it did), but I adjusted and moved forward, switching him to formula. For the record, Austin was always big, consistently landing in the 75th–90th percentiles for height and weight. I'm not advocating one feeding method over another—just sharing my experience. Whether it's breast milk or formula, you do what you must do.

So, when I noticed a slight drop in my milk supply while nursing Calla, I thought, Shoot, it's happening again. That's when I stumbled across the random Facebook post that introduced me to posterior tongue ties, upper lip ties, and everything they entail.

My first step was to make an appointment with an IBCLC (International Board-Certified Lactation Consultant). I wanted to see if there was anything I could do to help Calla breastfeed longer, and I was open to learning. Little did I know this appointment would prove pivotal, leading to a cascade of unexpected changes for our entire family.

While Austin was at Mother's Day Out, I took Calla to see Jamie, an IBCLC. The first thing Jamie did was weigh Calla, and then we sat down for a nursing session. Jamie carefully observed Calla's breastfeeding mechanics—how her tongue moved, the shape of her mouth, and her overall behavior while nursing. Afterward, we weighed Calla again to determine how much milk she had transferred during the session. Then Jamie conducted a hands-on exam.

She showed me the distinct dip in the middle of Calla's tongue and demonstrated its limited movement with a gloved pinky finger. Calla was diagnosed with a moderate posterior tongue tie and an upper lip tie. Jamie explained that Calla's latch was terrible. With my overproduction of milk, we had managed to maintain a working breastfeeding relationship. Jamie was genuinely surprised we had been able to nurse successfully for so long.

The next part of the exam involved Calla lying on her back while Jamie gently turned her head from side to side. Curious, I asked what she was looking for, and she explained she was checking for range of motion and neck tension. That's when I told her about my dad working on Calla at just a day and a half old, and again at three months. I confidently added that she shouldn't have any neck tension—but to my surprise, she did.

As Jamie turned Calla's head to the left and right, her opposite hip would lift. I was stunned. Jamie explained that when a baby has a tongue tie, they compensate for restricted tongue movement by engaging other muscles in their mouth and neck, even in utero, as they practice swallowing. My mind started racing with the implications of this, but then I mentioned something about Austin, and our conversation shifted in a new direction.

We discussed everything—from Austin's current eating habits to his short stint breastfeeding and his newborn gassiness. Based on what I shared, Jamie suspected that Austin also had a tongue tie and suggested that his restriction might be worse than Calla's.

Jamie also shared something that helped alleviate some of the lingering mom guilt I had over not nursing Austin longer. She explained that a mother's body produces breast milk based on hormonal control for the first three months after birth. After that, milk production is regulated by supply and demand. Because of Austin's suspected tongue and upper lip tie, he had no proper latch or suction, and therefore, no real demand. He wasn't sucking—he was merely drinking—while he was nursing. My milk supply dried up so quickly not because of the grieving and stress of losing my grandmother, but because of his tongue tie. That realization was unexpectedly relieving.

I also asked Jamie if a posterior tongue tie could be connected to speech delay. Austin, at over two years old, still wasn't talking. He communicated through grunts, gestures, sign language, and attempts at sounds, and I had become skilled at interpreting what he needed. Jamie explained that if the tie were severe enough, it could indeed impact speech. At the time, Austin was receiving speech therapy for his significant delay. That two-hour appointment turned out to be one of the most informative experiences of my life.

As I left the hospital, I called Dad, my mind racing as I tried to connect all the dots. The conversation went something like this:

Me: "Dad! Calla's neck is tight! It's because of a tongue tie! Her tongue doesn't move the way it's supposed to, so she's using other muscles in her mouth and neck to compensate!"

Dad: "WHAT?!"

Me: "Wait, Dad! Think about this—migraines! Jess and I have had migraines forever! Jess had an anterior tie released when she was little because of her lisp, remember? But she still has a posterior tie! And Jamie checked my tongue and said I have one too!"

We talked the entire way home, trying to process how all this information might be connected. When I picked Austin up from Mother's Day Out, I checked his shoulders and neck—they were rock hard on my two-year-old. He was so sensitive that he physically pushed my hands away because it hurt to be touched.

My next step was to consult our pediatrician, who, as it turned out, knew Jamie and fully trusted her recommendation to have the kids' tongue and lip ties revised.

There was a local dentist in the area who performed laser revisions for children under six months old without the need for anesthesia. Since Calla was just over six months old and Austin also needed a revision, I scheduled a consultation with our ENT. During their surgery, both kids were under general anesthesia for only 10–15 minutes each. As Jamie had predicted, it was confirmed that Austin had a severe posterior tongue tie and an upper lip tie.

After surgery, Calla nursed immediately, and her latch showed a noticeable improvement. She continued to breast-feed for just over two years, far longer than I ever anticipated. Within a week of the revisions, Austin began to articulate more words, and from that point on, his vocabulary exploded.

But our efforts didn't stop there. Both kids needed bodywork to release their tight muscles, so we incorporated massage (naturally), chiropractic adjustments, and craniosacral therapy into their care. While tongue ties may not be an obvious reason for pain in a child, the muscular shortening and tightening that can occur in the mouth and neck because of the tongue tie is very real.

An important note here is that our pediatrician was genuinely excited to discuss this topic during our referral to the ENT (all of this occurred in 2014). He mentioned that he was finishing a continuing education course but planned to review recent studies the following week—studies specifically focused on posterior tongue ties and their connection to breastfeeding. I deeply appreciated his openness and willingness to learn something that was not yet included in standard pediatric teaching.

Tongue Ties: A Hidden Link to Lifelong Issues

Our understanding of how tongue ties relate to conditions like migraines, TMJ, and other chronic issues comes from two things: hands-on experience and a fundamental understanding of how the body works as a unified system. Like a well-oiled machine, when one part is out of sync, it can disrupt everything around it.

As Dad often says, you can't fix a problem by only treating the symptoms. You have to get to the root cause.

After both kids had their tongue ties corrected, Tom and I decided to get checked as well. Tom had struggled with TMJ pain for years. I had lived with intense monthly migraines. After my procedure, the first thing I noticed was how much easier it was to swallow water. Something so basic, yet it had been affected the whole time without me realizing it. That moment gave me even more gratitude that we had caught and corrected the issue early for our kids.

Over the years, research on tongue ties has grown. Earlier studies, mostly from the mid-20th century, focused on how tongue ties affected breastfeeding and speech development in young children. But back then, almost no one explored the long-term impact.

In the early 2000s, more providers began recognizing the effects of posterior tongue ties. They noticed how restricted tongue movement forced babies to compensate with other muscles. By the 2010s, the conversation started shifting again, this time toward adults. We began to see how untreated tongue ties could contribute to jaw tension, TMJ, headaches, restricted airways, poor posture, and even chronic dental problems.

Now, with more adults seeking answers for symptoms that have long gone unexplained, we're finally starting to understand the full picture. Tongue ties are not just a childhood issue. They can shape a person's health for life.

But recognizing the symptoms isn't enough. We need to name what is really happening. And that starts with understanding PBPD.

CHAPTER FOUR

PRE-BIRTH POSTURAL DEFICITS (PBPD) IMPACT

Everywhere I looked, I saw the same story unfolding.

Parents would come to me with a baby who cried endlessly, did not sleep, hated diaper changes, or refused to nurse on one side or the other. Doctors had ruled out infections, allergies, and reflux. Some had mentioned colic. A few offered prescriptions. None could explain why their baby was miserable.

I applied the same method we use to figure out what is causing my grown clients' bodies to be unbalanced.

What I found were tight, over-contracted muscle groups. Patterns of tension that mirrored how that baby had been folded, twisted, or compressed in the womb. Fascia that had not fully unwound after birth. Muscles locked in postures that made normal movement painful.

I started seeing it everywhere.

These were not isolated cases. They were not random or unexplainable. They were deeply consistent. And yet, no one had given it a name.

So I did.

THE OVERLOOKED AND UNDIAGNOSED SOURCE OF INFANT STRUGGLES

Pre-Birth Postural Deficits (PBPD) refer to the muscular and fascial tension patterns in newborns that develop from how they were positioned in the womb, especially during those final, cramped weeks of pregnancy.

PBPD are not a disease. They are not a formal diagnosis yet. You will not find them in a pediatric textbook. But they are very real. I have seen these patterns in hundreds of babies.

Simply put, the way your baby was shaped in the womb can have a profound impact on how they function after birth.

For example, tightness in a single hip flexor can make diaper changes painful. Shortened neck muscles can cause torticollis. Deep restrictions in the chest or abdomen can lead to reflux, gas, or disrupted sleep. These are not random or mysterious symptoms. They are predictable outcomes of unresolved tension.

PBPD are often at the core of what many parents and even doctors describe as "normal baby problems." But normal is certainly not optimal. The good news is, these issues are almost always treatable.

Every baby carries some degree of tightness from the womb. None of us come out with perfect function. This is a universal reality, not an exception. And it is why babies need help.

The consequences of leaving PBPD unaddressed can ripple through every stage of development. But the opportunity here is incredible. When we recognize and treat these restrictions early, ideally right after birth, we give children the chance to grow, move, and thrive without the discomfort that so many families have come to accept as inevitable.

Let us look closely at three foundational areas of development: cognitive, physical, and social. Each can be dramatically enhanced by simply removing pain and restoring freedom of movement in a baby's body.

COGNITIVE DEVELOPMENT AND THE COST OF EARLY PAIN

A baby's brain develops more rapidly during the first two years of life than at any other time. This window is when the brain forms more neural connections than it ever will again, laying the foundation for a lifetime of learning and adaptation. During this critical period, infants use all their senses to explore the world around them, forming the synapses that will shape their intellect and cognitive potential (Harvard Center on the Developing Child).

These early neural pathways become the scaffolding upon which all future knowledge is built. But here is a simple question for you, the adult reading this: How much do you really learn when you are in pain?

Probably not much. That is because the human brain is wired to prioritize pain. It is a survival response that diverts attention and energy to protect us. Even as adults, when we attempt to function while in pain, our productivity and focus usually suffer.

Now imagine an infant, one who has no job to do, no tasks to complete, and nothing else to occupy their attention. If they are in constant physical discomfort, that pain becomes their entire world. Most babies in pain will not coo, smile, or play unless something temporarily distracts them. Their brains, for the vast majority of their first two years, are not building bridges for language or creativity. They are simply trying to cope.

And yet, that same vulnerability presents an extraordinary opportunity.

What if your child or grandchild never had to focus on pain in their earliest days? What if, instead of managing discomfort, their minds were free to explore language, shapes, colors, sounds, music, touch, energy, connection, and peace? What kind of neural pathways might form when a child's full attention is available for curiosity, not for coping?

How might their cognitive development flourish if it were not burdened by constant pain signals?

While this idea was once only theoretical, since it is likely no infant has ever grown up completely free from the tension and restriction caused by PBPD, we now have evidence to support it.

Recent clinical studies confirm that persistent pain in infancy can directly interfere with brain development. When pain is untreated or inadequately managed during this early stage, it disrupts neural processes responsible for learning, memory, and emotional regulation.

According to a review published in Pain, early exposure to chronic pain alters brain development and can have long-term effects on intellectual function, behavior, and stress response systems.[*]

In short, this is not just about comfort. It is about opportunity.

Helping babies out of pain is not only an act of compassion. It is an investment in their future.

DISRUPTING THE BRAIN'S GROWTH PATHWAYS

In the earliest months of life, the brain is undergoing rapid development, forming new connections at an extraordinary pace. During this critical time, persistent pain activates the hypothalamic–pituitary–adrenal (HPA) axis, prompting the release of stress hormones such as cortisol.

Excessive exposure to cortisol can interfere with the development of key brain areas, including the prefrontal cortex, which governs decision-making and problem-solving, and the hippocampus, which is essential for memory and learning.

LIMITING NEUROPLASTICITY

Neuroplasticity, the brain's ability to adapt, reorganize, and form new connections, is especially active during infancy. However, repeated exposure to pain can hinder this process. The brain becomes less capable of forming the flexible, resilient networks required for intellectual growth and emotional recovery from stress.

---

[*] Walker, S. M. (2019). Translational studies identify long-term impact of prior neonatal pain experience. Pain, 160 (Suppl 1), S1–S10. https://doi.org/10.1097/j.pain.0000000000001551

COGNITIVE AND DEVELOPMENTAL CONSEQUENCES

Untreated pain during infancy has been associated with several concerning outcomes, including:

- Lower scores on IQ and cognitive assessments
- Delays in speech and language development
- Challenges with attention, executive functioning, and problem-solving

LONG-TERM EMOTIONAL AND BEHAVIORAL STRUGGLES

Ongoing discomfort in early life is also linked to higher levels of anxiety, poor emotional regulation, and increased sensitivity to stress. These challenges can follow a child into school years, compounding learning difficulties and impacting academic achievement.

WHAT THE RESEARCH SHOWS

A study published in Restorative Neurology and Neuroscience found that repeated painful experiences in neonates are strongly associated with delays in cognitive and motor development. Infants who underwent multiple medical procedures without adequate pain management later scored lower on attention and memory tests in early childhood. *

---

* Mansour, A. R., Farmer, M. A., Baliki, M. N., & Apkarian, A. V. (2014). Chronic pain: the role of learning and brain plasticity. Restorative Neurology and Neuroscience, 32(1), 129–139. https://doi.org/10.3233/RNN-139003

Another study in Frontiers in Pediatrics emphasized that early life stress, including unmanaged pain, can adversely shape brain architecture. Encouragingly, it also found that interventions such as skin-to-skin contact significantly improved emotional and cognitive outcomes by lowering stress levels and supporting healthy neural development.[*]

HOW MOENCH METHOD BODYWORK® ACCELERATES HEALTHY BRAIN GROWTH

Moench Method Bodywork® provides a powerful solution to these risks by identifying and relieving sources of pain that might otherwise be missed. Through gentle, targeted interventions, this approach supports optimal brain development in three key ways:

- Restoring healthy neural function: By alleviating pain, the brain is no longer preoccupied by discomfort and can return to building normal, adaptive connections.
- Enhancing cognitive capacity: Infants who are no longer consumed by pain are better positioned to develop memory, focus, and problem-solving skills.
- Promoting emotional balance: When the stress response is no longer being triggered by pain, infants can begin to develop the self-regulation skills necessary for healthy emotional and social development.

---

[*] Smith, K. E., & Pollak, S. D. (2020). Early life stress and development: potential mechanisms for adverse outcomes. Journal of Neurodevelopmental Disorders, 12, 34. https://doi.org/10.1186/s11689-020-09337-y

THE STAKES COULD NOT BE HIGHER

Pain in infancy is not a phase to be ignored. It has the power to diminish how a child learns, feels, and connects with others for years to come. Left untreated, it becomes a barrier to growth and learning. But when pain is addressed early and effectively, we unlock a child's natural capacity to thrive.

One of the primary reasons Moench Method Bodywork® exists is to remove that barrier. Every child deserves the chance to grow free from pain. When we remove that burden, the possibilities for their future become limitless.

Let us not waste that opportunity.

PHYSICAL DEVELOPMENT AND THE IMPACT OF EARLY PAIN

Of all the ways pain affects a child, its impact on physical development may be the easiest for adults to understand. Think about what you do when your leg hurts after raising it too high. You stop raising it that way. Your body naturally avoids the movement that causes discomfort. Pain is a warning system. It signals when something is too tight and at risk of injury.

This instinct is not limited to adults. Babies respond the same way. When a movement causes pain, they remember it. The next time, they avoid that movement or modify it. This early lesson, do not repeat what hurts, stays with us for life.

I saw this play out clearly when my oldest daughter, Lauren, was a toddler. She was walking into her aunt's room for the first time, stepping over a low threshold where the garage had been converted. She did not expect the change in elevation and suddenly face-planted on the carpet. A few days later, she approached her grandpa's room, a normal, flush doorway, not the step down of her aunt's room. As she got close, she dropped to all fours, turned around, and backed over the threshold with extreme caution. She had learned her lesson. Pain had taught her to adapt.

This story may be lighthearted, but it illustrates a deeper truth. Many of our lifelong movement patterns begin this way, as subtle adjustments to avoid pain. These compensations become habits. Over time, they affect how we walk, run, sleep, and carry ourselves.

Now imagine a baby dealing with tension from pre-birth postural deficits. How much taller, stronger, faster, or more fluid could that child be if their body was not compensating for pain they had no words to describe?

While we cannot know that answer precisely, my decades of experience tell me the potential is massive. I have worked with clients since the 1990s, many of whom struggled with issues rooted in early muscular and fascial restrictions. Many of our adult clients still carry pre-birth postural deficits. As the weight of their head and torso gradually shortens anterior neck muscles and hip flexors, this strains the lower back and gluteal muscles until they are forced to hunch forward.

Layer on old injuries, poor training habits, and sedentary routines, and the restrictions compound. All of it is proof that an early tune-up could have spared them years of daily limitation and pain.

The list of common problems that could be prevented, or at least minimized, is long. It includes scoliosis, torticollis, hip dysplasia, leg length discrepancies, chronic back and neck pain, shoulder tightness, migraines, herniated discs, poor posture, panic attacks, carpal tunnel syndrome, knee and hip pain, and even joint degeneration. The real question is whether the medical community is ready to address the root cause of these problems or whether they prefer to continue managing symptoms.

To put this in perspective, consider what happens when left-handed children are forced to become right-handed, a practice that was once common. Decades of research now show that such intervention can disrupt neurological, emotional, cognitive, and motor development. For example, positron emission tomography (PET) studies have found that adults who were forced to switch hand dominance as children show persistent differences in brain activity (Siebner H. R. et al., Journal of Neuroscience, 2002). These individuals often experience increased rates of dyslexia, speech disorders such as stuttering, and behavioral challenges.

Emotionally, they may become more anxious, withdrawn, or prone to neurotic behaviors. Their fine motor skills also suffer, making everyday tasks more difficult than they should be. These effects reveal a broader principle: when we alter a child's natural movement patterns, whether through hand dominance or compensatory posture, we can create lasting developmental consequences.

We see the same effects when someone favors one side of their body due to pain, injury, or even sleep position. These imbalances prompt the brain to adapt, rewiring how muscles fire and move. At first, the changes may seem harmless. But over time, they lead to reduced mobility, joint strain, chronic pain, and systemic health issues.

This is not just theory. Research backs it up. A study from Harvard Medical School, part of the Football Players Health Study, examined 3,506 former NFL athletes between the ages of 39 and 67. The data showed that players who had undergone ACL surgery had a 50 percent higher risk of heart attacks. They were also twice as likely to need knee replacements and had a 50 percent greater risk of developing arthritis.

Why does this happen? One study suggests that reduced physical activity following injury, along with chronic low-grade inflammation, contributes significantly to the increased cardiovascular risk (Mathur N., Pedersen B. K., Mediators of Inflammation, 2008). Although these studies do not claim direct causation, they point to a clear link: when injuries, imbalances, or early movement restrictions are not fully addressed, the body adapts in ways that create long-term health consequences.

What begins as a simple compensation, like favoring one leg or avoiding a painful movement, can snowball into joint degeneration, chronic inflammation, or even heart disease. These findings underscore a powerful truth: you cannot simply work around pain, imbalance, or a pre-birth postural restriction forever. Sooner or later, it catches up.

I believe that the tightness and restriction of the soft tissue in the lower extremity of the injured athlete impedes proper blood supply and lymphatic drainage to and from that limb. That causes an increase in blood pressure throughout the body and adds significant stress to the heart and every other system in the body, which results in the conditions listed above.

SOCIAL DEVELOPMENT: THE FIRST BONDS BEGIN IN THE BODY

Human connection starts before we can speak. It begins with eye contact, physical touch, facial recognition, and shared rhythms like rocking, holding, and soothing. These early social interactions form the foundation of every relationship a child will have.

But what happens when touch causes pain instead of comfort?

When a baby flinches, cries, or stiffens every time they are picked up, held, or changed, something important is lost. Not just physical ease, but trust, bonding, and emotional safety. These moments of withdrawal, repeated dozens or even hundreds of times a week, start to shape the infant's sense of the world. Is it safe? Is it warm? Can I relax into the arms that hold me?

Infants experiencing persistent discomfort, whether from tight fascia, joint misalignment, or muscular tension, often avoid touch without knowing why. And parents, sensing this aversion but not understanding it, may pull back as well. The result is a silent fracture in what should be a joyful, nurturing connection.

Over time, this physical disconnection can evolve into emotional distance. Children who experienced pain during their earliest interactions may struggle to bond, self-regulate, or connect socially. They may avoid eye contact, resist cuddling, or have difficulty expressing or understanding emotions. These early disruptions in sensory and emotional development can ripple into anxiety, isolation, or behavioral challenges later in life.

Scientific research supports this link between early pain and altered social development. A study in Neuroscience and Biobehavioral Reviews found that early exposure to stress and pain can alter a child's attachment behaviors and stress-response systems (Grunau R. E. et al., 2007). These changes may affect not just emotional health but also learning, social confidence, and resilience.

Thankfully, the reverse is also true.

When a baby's pain is resolved, when their body is free to move, stretch, and receive touch without distress, their world begins to open. They can bond more easily. They look at faces longer. They begin to smile, explore, and engage. As they feel safer in their body, they feel safer with others.

Moench Method Bodywork® is not just about relieving pain. It is about restoring trust, between a baby and their body, and between a child and the world.

When touch feels good instead of threatening, everything changes. Connection deepens. Joy bubbles up. A foundation is built for healthy relationships, emotional intelligence, and a lifelong sense of safety in the world.

This is what every child deserves.

The earlier we spot these patterns, the better the outcomes. But to do that, we need a different way of seeing.

WHAT WE NEED TO UNDERSTAND TO HELP OUR BABIES THRIVE

I have spent more than thirty years reading bodies, but the moment I was doing research and saw a newborn's pain scan, I felt the air leave the room. A single heel stick lights up 18 of the 20 brain regions adults use to register pain, about a tenth of the cortex pulled away from the construction crew that is supposed to be wiring vision, language, and social connection.

That alarm does not ring just once. In the NICU, a preterm baby is poked, prodded, or squeezed roughly 14 times a day during those first fragile weeks. The more hits a child takes, the steeper the brain tax. What happens to all these babies experiencing the conditions we mentioned earlier?

MRI work shows that heavy early pain exposure leaves the brain's switchboard, the lateral thalamus, 4 to 6 percent smaller and the wiring to the thinking cortex sluggish enough to predict lower memory and attention scores years later.

Here is the part that keeps me hopeful: when we shut down pain, the brain rushes to claim its stolen territory. When we add targeted hands-on work to ease the silent muscle tension I so often find after birth, and that reclaimed 15 to 25 percent of processing power no longer has to manage pain, it can drastically increase eye contact, curiosity, and the lightning-fast learning every baby is wired for.

CHAPTER FIVE

A DIFFERENT WAY OF SEEING

HOW TO READ THE SIGNS—WHAT YOUR BABY'S BODY IS TRYING TO TELL YOU

We are not trained to see pain in babies. When you recognize that your baby's cry isn't a mystery, but a message, you stop feeling helpless. You become their first healer.

We are trained to look for symptoms: crying, arching, feeding issues. Then we manage those symptoms with noise machines, new bottles, gas drops, or by "waiting it out." But when you begin seeing the body differently, you realize the baby is already telling you exactly what is wrong. You just need to know how to look.

That is what this chapter is about. Not treatment. Not diagnosis. Just observation. It is the difference between being a parent who is at the mercy of "experts" and being one who is equipped to see what most people miss.

Let's start with the basics.

TIGHTNESS TELLS THE TRUTH

When adults are in pain, we say so. We describe it, or we show it by limping, protecting the area, or asking for help. But a baby has none of those tools. They cannot say "my neck hurts" or "my hip feels jammed." So instead, they:

- Arch their back while crying
- Only nurse comfortably on one side
- Get stiff during diaper changes
- Clench their fists or curl their toes
- Seem easily startled or touch-averse
- Struggle to sleep unless swaddled tightly

Each of these behaviors is a clue. And many of them point to the same root issue: muscle and fascia tension that never released after birth.

COMMON CLUES TO WATCH FOR

Let's get specific. These are some of the most common signs I see in infants with PBPDs:

- Asymmetrical leg movement during diaper changes (one leg resists)
- Turning the head only one direction while lying down or nursing
- A flat spot on one side of the head (often from not turning both ways)
- Crying during tummy time or avoiding it altogether
- Excessive startle reflex, even in calm environments
- Discomfort with touch — the baby stiffens, squirms, or cries when held a certain way

These are not random quirks. They are body language. They are your baby's way of saying, "Something hurts. Please help me."

SEEING THROUGH THE NOISE

It is easy to miss these signs because we have been told they are normal, or that they will go away on their own. Sometimes they do. But often, the baby just adapts. They adjust to the restriction. They figure out how to live with the discomfort. And that adaptation becomes their baseline.

Here is the problem: adaptations come with a cost.

A baby who cannot turn their head fully may develop a visual tracking delay. One who avoids tummy time may crawl late or not at all. One with a tight jaw or neck might struggle to speak clearly later or deal with migraines as a teen or adult.

The earlier you can spot the pattern, the better the outcomes.

What looks like "just how they are" can actually be a lifelong workaround. A baby who compensates now may become a child who moves awkwardly, a teen who slouches to avoid discomfort, or an adult who has no idea why they always feel "off."

A PARENT'S SUPERPOWER

You don't need a degree to see your baby clearly. You just need a new lens and the courage to trust what you already feel. You already know your baby better than anyone else. What this chapter gives you is a way of interpreting what you are seeing and what your baby might be feeling.

That is your superpower. And it is the beginning of real healing.

In the next chapter, I will walk you through exactly what our treatment process looks like. You will see how simple movements, applied with knowledge and care, can relieve years of tension in minutes. But none of that matters if you do not first learn to see.

CHAPTER SIX
THE TREATMENT PROCESS
WHAT IT LOOKS LIKE AND WHY IT
WORKS

STEP-BY-STEP INSIGHTS INTO HANDS-ON THERAPY

Let's get practical.

You now understand that many babies hurt, and that pain often begins before they are even born. You've seen how tight fascia and muscle restrictions can show up in colic, torticollis, feeding issues, poor sleep, and developmental delays. But now you may be asking the most important question of all: **What can we actually do about it?** This chapter will show you.

WHAT IS THE TREATMENT, EXACTLY?

At its core, the treatment is hands-on fascia and soft tissue release. It is not a typical massage, not chiropractic, and not a physical therapy session. Think of it as highly skilled, deeply informed bodywork designed to gently but precisely restore freedom of movement in areas that have been restricted for months, possibly years.

Treatments are done with the baby lying on a table. The practitioner uses gentle pressure, typically with fingers and sometimes the palm or forearm, to engage a muscle group or fascial line. We are not poking, jabbing, or adjusting. We are listening. Feeling for tension. Following the body's signals.

Sometimes it takes just minutes for a muscle to release. Sometimes it takes longer. But in almost every case, the change is immediate and noticeable.

WHAT DOES IT LOOK LIKE DURING A SESSION?

Here is a basic walkthrough of a typical infant session:

1. Observation

We begin by watching how the baby holds themselves. Do they tilt their head? Pull one leg higher? Avoid turning to one side? We ask parents what they've noticed about how the baby moves or holds themselves, and which positions seem more comfortable for them.

2. Palpation and Light Testing

We gently test for muscle tone, tightness, and tenderness. We check the hips, shoulders, back, neck, jaw, and more.

3. Targeted Release

Using a fingertip or thumb, we apply sustained pressure into a tight muscle group. If the hip flexor is shortened, for instance, we might gently bend and straighten the leg while holding pressure.

4. Response Monitoring

Babies often cry during release. Not because they are being harmed, but because it is the only way their nervous system can express discomfort. The crying often changes in quality: it starts sharp, then softens, and finally stops with a deep exhale or sigh.

5. Integration and Rest

After a release, we often see the baby stretch out, fall asleep, or exhibit calm alertness. This is how we know their nervous system has shifted into healing mode.

WHAT DOES IT FEEL LIKE FOR THE BABY?

Most releases are uncomfortable. But discomfort is not the same as danger. This work is no more painful than the ache you feel during a deep stretch or after a tough workout.

Babies often express this through crying. As a parent, you will learn to hear the difference: the cry of distress versus the cry of release. The moment the breath softens, the moment the body goes limp, you will know what just happened. It is the body letting go of something it has held for far too long.

And once it's gone, everything changes.

WHY IT WORKS

Muscle and fascia do not stretch out on their own after birth. If they remain tight, they interfere with posture, digestion, breathing, even emotional regulation. When you apply precise, intentional therapy to the right areas, the body responds. It unwinds. It reorganizes itself.

Pain disappears because the cause has been removed, not masked.

This is not magic. It is mechanics. And once you've seen it, it feels obvious.

A WORD ABOUT CRYING

No one likes to hear a baby cry. Least of all a parent who already feels helpless. But in our treatment rooms, crying often signals progress. It means we found the spot. We are working through something real.

And when the crying stops — not because the session ended, but because the tension finally released — many parents cry too. From relief. From hope. From finally seeing their child at peace.

WHAT TO EXPECT AFTER TREATMENT

Many babies sleep deeply after their first session. Some experience temporary fatigue or fussiness as their body adjusts. But almost all exhibit noticeable changes:

- More relaxed posture
- Improved sleep
- Easier feeding
- Less crying
- Increased movement and coordination

For some babies, a single session makes a dramatic difference. For others, it may take two or three to unwind all the holding patterns. But the changes stick because they are rooted in the body's structure, not imposed from the outside.

YOU DON'T HAVE TO BE AN EXPERT TO SEE IT

Most of our clients come in skeptical. They leave amazed. Why? Because they watch it happen.

You will see your baby's body change. You will feel the shift in your arms. And you will know — without needing a clinical trial — that something real just happened.

In the next chapter, we'll explore how these same patterns show up in teens and adults. Because this work is not just about babies. It is about everybody who has ever learned to live with pain.

CASE STUDY: HIP AND NECK ON A DANCER

Several years ago, one of my longtime clients, a college professor named Harvey, referred one of his students to me. She was a classical ballet major who had been struggling with persistent hip pain. Her discomfort was beginning to interfere with both her training and performance.

The cause turned out to be relatively simple. She slept on her side every night, which caused one leg to rest higher than the other. Over time, this habit created tension in her hip flexor. I released the muscle, and just like that, her hip pain and restricted range of motion were gone.

But during our session, I noticed something else. She seemed to have difficulty turning her head to one side. When I asked her about it, she replied, "It's always been like that. In every photo from childhood, my head is tilted or turned that way."

I told her we would take a closer look during her next appointment.

At that follow-up session, I worked through the deep, tight muscles in her neck that were limiting her range of motion. It was not a comfortable process, but the result was dramatic. When she got off the table, she turned her head fully, pain-free, for the first time in her life.

She looked left, paused, and said something I will never forget:

"I've never seen my left shoulder before."

It was a powerful moment. It also came with an unexpected twist.

When she returned to dance class, she kept stumbling. Turning her head in a new way had shifted her balance. Her center of gravity was off, and her body had to relearn its orientation in space. Fortunately, her instructor understood what was happening and encouraged her to stick with it. Within a week, she had adjusted to her new alignment and was dancing beautifully.

CASE STUDY: TORTICOLLIS IN A 3-YEAR-OLD

Not long after working with the dancer, I was treating a firefighter who listened intently as I shared her story. He paused, then told me about his three-year-old daughter, Clara. Her head had a noticeable tilt to one side, and her doctors were considering eye surgery. Her vision had begun adjusting to her tilted horizon.

The official diagnosis was torticollis. This was before we regularly worked on babies.

I researched it. It was classified as a congenital condition — something present at birth — with no known cause and no definitive cure. To me, it sounded like she had a tight, sore neck, possibly from her position in the womb or a restriction from birth.

We scheduled some sessions to see what might help.

Sure enough, after just five or six treatments, each lasting about ten minutes, Clara regained full range of motion in her neck. The pain was gone. She could move freely, look straight ahead, and turn in both directions without strain.

When I last checked in with her dad eight years later, Clara was still completely pain-free.

Below is what Clara's mother, Anna, shared about their experience:

"My daughter, Clara, was born in February 2004… We went to physical therapy at ten months, but it did little to correct the problem… After seeing many specialists, including neurologists and orthopedists,

no one could help. Ken treated her six times, ten minutes each, and her neck went from a 20-degree tilt to a 5-degree tilt. The results were immediate… After additional care, including chiropractic support, her neck is now completely straight. I will forever be grateful. My only regret is that doctors didn't know about this treatment when she was an infant."

CASE STUDY: "SPORTS-INDUCED ASTHMA"

Another young client, a 14-year-old athlete, came in complaining of lower back pain. Over several months, she experienced a series of sports-related injuries. At first, her case seemed routine.

But during the session, her mom mentioned something else: her daughter had recently begun struggling to catch her breath while running. The doctor diagnosed her with exercise-induced asthma and prescribed an inhaler.

They tried it three times. Nothing changed.

While reviewing her chart, I asked the girl, "Are you still sleeping on your side?"

"Yes," she said. "I try to sleep on my back, but I always end up on my side."

Her mom added, "Just like she did when she was a baby."

That was the missing piece.

There is a group of muscles near the top of the rib cage that help lift the chest during deep breathing — one of the main ones is the serratus anterior, which attaches to the first eight ribs. These muscles often become shortened when someone sleeps on their side. Contracted and compressed, they struggle to expand during breathing.

If those muscles cannot lift the chest fully, the result is shallow breathing. In her case, this led to a mistaken asthma diagnosis.

I have seen similar patterns in adults with panic attacks who were prescribed psychiatric medication, even though the root issue was muscular tension caused by sleep posture.

There was also a clear connection between her breathing challenges and her injuries. Shallow breathing reduces oxygen delivery, leading to faster fatigue, more lactic acid buildup, and a higher risk of injury.

Every one of her issues — injuries and "asthma" — traced back to a simple root cause: her sleep posture. A pattern she developed as an infant and carried through adolescence.

Although most visible in babies, these same patterns appear in teens, adults, and even seniors. They simply show up in ways we don't always think to connect.

CHAPTER SEVEN

THE LONG SHADOW OF UNRESOLVED PAIN

The truth is, this work is not just about babies.

Yes, the most urgent window for change is in infancy, when soft tissue is pliable, patterns are still forming, and pain is often expressed in cries that no one can decode. But what happens if no one ever decodes them?

What happens when a tight hip flexor goes unnoticed? When a baby never stretches their neck fully? When shallow breathing becomes the norm because of early chest compression? The answer is that the body adapts, but at a cost.

The toddlers who never quite crawl right become the kids who trip, fidget, or sit awkwardly in their chairs. The kids who hate being touched, often called "tactile defensive," grow up craving connection but recoiling from it. The teenagers with chronic headaches, jaw tension, or back pain do not know that what they are carrying might have started before their first birthday.

And adults? They have lived with it so long, they just call it "normal."

That is why this chapter exists. To help you see the through-line. To connect the dots between a baby who cries during diaper changes and a woman who misses work every month from menstrual pain. To understand how one untreated restriction in the womb can echo forward for decades.

I have treated adults in their fifties who finally, for the first time, understand why their body never quite worked the way it should have. And I have watched their posture, breath, sleep, and energy change in days. Not because of medication. Not because of surgery. But because we addressed something that had been hiding in plain sight for a lifetime.

So the real question is this: if pain and dysfunction can start this early and persist this long, why have not more doctors caught on?

What Do Doctors "Know" About Pain or Soft Tissue Problems?

They know what they were taught.

The reality is this: most doctors have little to no under-standing of why so many people are in pain.[*] Think about it. How many physicians do you know who specialize in soft tissue? Probably none. One important exception is the physia-trist. These doctors are trained to treat the whole body, not just isolated symptoms, and they understand the role of soft

---

[*] https://www.nytimes.com/2025/01/12/magazine/chronic-pain.html

tissue in pain and dysfunction. It is a rare, refreshing approach.

Here is the irony. Fascia and other soft tissues are rich with pain receptors, yet the standard medical curriculum devotes little time to hands-on release techniques. It is not a lack of care or skill on the part of physicians. It is a system designed to reward procedures and prescriptions more readily than root-cause resolution. When financial and time incentives favor symptom management, true healing can be crowded out. The price we pay is not only economic; it is physical, emotional, and ultimately generational.

Take my baby brother. He was born with legs that did not align properly. That is common for babies whose position in the womb compresses the lower body. The standard medical response was braces. He spent months wearing orthopedic shoes connected by a metal bar at night, just like a young Forrest Gump. If someone had simply released the tight soft tissue, he could have skipped all of it.

These devices are meant to "train" the muscles and fascia by force. But unless there is a true skeletal abnormality, that same realignment can often be achieved much faster, and far more gently, by manually lengthening the tissue instead.

The same applies to infants born with clubfoot. The traditional treatment is called serial casting. It is a drawn-out, rigid process that uses cast after cast to force the foot into position. It immobilizes the leg for weeks, with each cast moving the tissue a little more. It is grueling, expensive, and slow.

What do we do? We release the soft tissue directly. No casts. No force. No delay. The body relaxes into the correction naturally because that is what the tissue wants to do when it is no longer locked down.

Doctors do what they were taught. In many cases, the procedures are decades old. Doctors still follow them because someone wrote a book and it was part of their medical school curriculum. That is not their fault. But it is still our problem.

For infants, our goal is simple but profound: give them the strongest start in life possible. I believe that each of us is here for a reason. And I believe that pain and dysfunction block us from living out our full potential. In my opinion, unrealized potential is one of the greatest tragedies there is.

If you want your child to sleep through the night…

If you want your child to grow up smarter, stronger, faster, more balanced, and more emotionally secure…

This is the bodywork they need.

EPILOGUE

I am genuinely honored that you chose to read this book, and even more so that you made it all the way through. The truth is, there is far more I do not know than I do. But there are two things I believe with absolute certainty: our bodies are built to be healthy and pain free, and most of the pain and dysfunction we carry has been with us since the womb.

I believe early child development remains one of the most overlooked frontiers in medicine. In many ways, the child development field is still in its infancy and will remain there until PBPD becomes part of the standard vocabulary.

We talk often about milestones and measurements, but we rarely talk about what becomes locked into the body before those first steps or first words. If we did, we would see that some of the most powerful opportunities to shape a child's future physically, emotionally, and cognitively begin by addressing the tension they are born carrying.

If this book has shown you something new, please share it. Share it with parents, grandparents, caregivers, and health professionals who are open hearted and open minded. The more people who know what to look for, the more babies we can help. And the more adults we might save from a lifetime of hidden pain.

My goal, believe it or not, is to work myself out of a job. I want this work to reach every practitioner who is willing to learn it. Too many people are hurting when they do not need to be.

Let's change that. Starting today.

We are developing online training courses specifically designed for treating infants. Please contact us if this is something you, or a practitioner you know would be interested in learning this life altering treatment technique.

Blessings and peace to you and your loved ones.

Kenneth R. Moench, LMT, MTI

P. S. This is my second attempt at writing my first book. Years ago, I had begun writing about my practice, how I developed the techniques we use, exactly what those techniques are, and how they differ from other methods used to treat pain, injuries, or physical dysfunctions. I had even asked my clients to bring me their stories and testimonials, along with a photo of themselves for the book.

Around that time, one of my clients, Wade, asked if I had a website and, if not, whether I would like him to put one together. I hesitated, then said, "Sure, I guess. If it can do X, Y, and Z, sure."

A little while later, Wade called and asked, "Do you have anything like who you are, what you do, how you do it, maybe some testimonials?"

"Sure," I replied, knowing full well that my book had now become my website.

Looking back, I am glad it happened that way. You are not really in business unless you have a website. Over the years, the site has grown to address the common questions people ask when they call the office and to provide information we want clients to review after their treatments.

The web address is **www.MoenchMethodBodywork.com**.

You will find a lot of information there, some of it practical, some of it humorous, and hopefully all of it helpful.

Enjoy!

ABOUT KENNETH R. MOENCH, LMT, MTI

Ken Moench (pronounced Mench) has spent more than thirty years turning chronic pain horror stories into comeback highlights.

After beginning his career as a physical therapy aide and later changing course to massage therapy, he developed the Moench Method, a precise blend of myofascial release, deep tissue work, trigger point therapy, and guided range of motion that addresses pain at its source instead of masking symptoms.

His four Central Texas clinics welcome everyone from newborns to world class athletes, and most clients need only two or three visits before they walk out pain free. Elite athletes from multiple professional sports trust the Moench Method to return to the field faster and perform at their peak. That same confidence led the founders of Austin FC to bring Moench Method Bodywork® in house for the entire roster, something no professional club had done before. Collegiate standouts follow suit, using NIL (Name, Image, and Likeness) sponsorships to cover their travel and treatment so they can compete pain free and protect their future careers.

Altogether, Ken and the therapists he has trained have helped tens of thousands of people ditch pills, avoid surgery, and return to living the lives they were meant to live.

In this book he shares insights about what may be the most overlooked root cause of nearly every ache, pain, or dysfunction: **Pre-Birth Postural Deficits.**

ACKNOWLEDGMENTS

This may be the most difficult section of this book to write because there are so many people to thank for helping bring it to life.

First and foremost, I want to thank my wife, Libby, who has put up with my long hours, late nights, weekends, and an exhausting travel schedule that takes me away from home far too often. She has edited countless versions of this book and handled the back end of my business for years without nearly enough thanks or fanfare. I can honestly say that without her help in the early days—on every aspect of this practice—we would not still be in business today. That is not an exaggeration in the least. Thank you so much, Libby. I love you, sweetie.

I want to thank all of the therapists I have trained throughout the years. Each of you has shed light on the complexities of the work we do every day. Your questions, your struggles, and your eventual mastery of this technique have been truly inspirational. I have learned so much *from* each one of you, and I hope you have benefited from our time spent helping people who felt everyone else had given up on them.

Special thanks go to two of my therapists whose impact has been exceptional for both the company and for me personally.

Jordan Daugherty has been with me for more than thirteen years. He started as a 23-year-old single guy with his own country band. Today he is married, the father of two daughters, and trusted by everyone from Austin FC players to elite athletes, grandparents, babies, and weekend warriors. He can handle any condition with professionalism, humor, and tremendous skill. Thank you, Jordan, for putting up with me all these years.

The second is Chris Nowels. He is one of the smartest, hardest working, most compassionate and caring people I have ever known. He has made countless invaluable changes and innovations that have helped move Moench Method Bodywork® to where it is today. He has challenged me, pushed me, and helped shape me into the leader and mentor I needed to become so this company could grow and accomplish its mission—to help alleviate unnecessary pain and dysfunction across the world. He is truly a good man and an incredibly talented therapist. From the bottom of my heart, thank you, sir.

I need to thank my firstborn daughter, Lauren, who assumed the role of C.O.O. and has been the backbone of Moench Method Bodywork® for several years. She answers phones when office staff are with clients, schedules countless people each day, runs interference for me, handles advertising and budgets, determines where I need to be for training and problem-solving, manages payroll, stays on top of deadlines for taxes and licensure, calms the most desperate callers who are in pain, and assures them we can help. She books all of my flights, rental cars, and hotels, and always has back-up plans if anything gets delayed. She has gone through numerous website revisions, adds content on the fly, and keeps our point-of-sale system updated. I call her every morning on the way to work and every evening on the way home to go over what still needs to be handled.

And while she is doing all of that—and much more—she is also homeschooling her two children, caring for three dogs, and being a wonderful wife to her husband, Tom (who is one of our biggest supporters because he knows how much this work has helped him, his family, and so many of his friends and co-workers).

If we logged all the hours Lauren has spent reading, editing, and re-editing *Every Baby Hurts*, nobody would believe it. She picked it up, dusted it off, and breathed life back into it. "We have to finish this book, Dad!" she said. A few months ago she pushed it over the finish line once again. Without her, this book would still be an unfinished manuscript on my laptop. *Thank you* feels wholly inadequate… and it is. I love you, sweetie.

I also want to thank a new friend, Bryan Eisenberg, whom I met at our Round Rock office's 25th Anniversary Ribbon Cutting Ceremony. After hearing me speak, he said, "I didn't get *any* of that from your website. This is amazing!" He told me I had a book in me. I told him I had a book that needed to be finished. Bryan is an accomplished, bestselling author, and I am not. He helped me expand key parts of this book, clarify its message, and shape it into what you are reading today. Bryan, thank you for your kindness, guidance, and encouragement.

Lastly, I want to thank the owners of Good Story, Tommy and Lisa Zukoski. They helped with the final edits of *Every Baby Hurts* and convinced me the original manuscript was actually two books, not one—and that separating them would help more people discover the truth in these pages. They are also updating our website, expanding our social media presence, and handling a host of other items they excel at. Thank you both, from the bottom of my heart.

Thank you to each and every one of you. This book exists because of your belief, your time, and your support.

AFTERWORD

This may be the most important part of this book, in my opinion.

I take no credit for the work I do or for the things I have learned about the human frame or how to ease suffering. At the beginning of this journey, I asked God specifically for knowledge, wisdom, and discernment—to show me a better, faster way to eliminate pain and dysfunction so that people could fulfill their purpose in this life. He has been patient with me for many years while continually answering that prayer.

Much later, I was reading one of my favorite authors, Leo Tolstoy. I was struck by his depth of knowledge and his seemingly limitless ability to translate that knowledge into written form. I remember thinking, *Everything must have already been thought of. Everything.* But then I wondered, *Maybe not everything.*

So I asked God if He would be so kind as to grant me one original thought.

"Every Baby Hurts" was that thought.

These two answered prayers are what made this book possible. It is proof to me that God uses the unqualified for His purposes. It is why I take no credit for any of it and why I give our Creator all of it. Thank you, Jesus.

"Call to me and I will answer you, and will tell you great and hidden things that you have not known."

Jeremiah 33:3

"Now to him who is able to do far more abundantly than all that we ask or think, according to the power at work within us."

Ephesians 3:20

"If you ask me anything in my name, I will do it."

John 14:14

Blessings to each and every one of you.

Ken

Ken's daughter, Stephanie, and son Lochlan age, 3 months.